BLACK MAMBA

by Craig Boutland

Consultant: Chris Mattison, herpetologist

BEARPORT
PUBLISHING

Minneapolis, Minnesota

Credits

Cover, ©NickEvansKZN/Shutterstock; 1, ©luk341/iStock; 3, ©tirc85/iStock; 4, ©Amit Rane/Dreamstime; 5, © NickEvansKZN/Shutterstock; 7, ©Avalon/Photoshot License/Alamy; 7B, ©Andrzej Kubik/ Shutterstock; 8, ©Peter Oosthuizen/Shutterstock; 9, ©Panther Media Gmbh/Alamy; 10, ©Andre Coetzer/ Shutterstock; 11, ©Michael D. Ken/Nature Picture Library; 12, ©Leopardinatree/iStock; 13, ©Karl van der Westhuizen/Shutterstock; 14, ©Adrian Warren/ARDEA; 15, © NickEvansKZN/Shutterstock; 16, ©Paul Freed/AnimalsAnimals; 17, © Photoshot/Maik Dobiey/Avalon; 18, © NickEvansKZN/Shutterstock; 19, ©scooperdigital/Shutterstock; 20, ©Bloomberg/Getty Images; 21, © NickEvansKZN/Shutterstock; 22TL, ©Marie Holding/iStock; 22CR, ©Robert Styppa/Shutterstock; 22B, ©Chris Mattison; 23, ©Eugene Troskie/Shutterstock; 24, ©Marie Holding/iStock.

T=Top, B=Bottom, L=Left, R=Right

President: Jen Jenson
Director of Product Development: Spencer Brinker
Editor: Allison Juda
Designer: Micah Edel
Editorial and Design: Brown Bear Books Ltd
Brown Bear Books has made every attempt to contact the copyright holders.
If anyone has any information please contact licensing@brownbearbooks.co.uk

Library of Congress Cataloging-in-Publication Data

Names: Boutland, Craig, author.
Title: Black mamba / by Craig Boutland.
Description: Minneapolis, Minnesota : Bearport Publishing Company, 2021] |
   Series: Slither! | Includes bibliographical references and index.
Identifiers: LCCN 2020008977 (print) | LCCN 2020008978 (ebook) | ISBN
   9781647470944 (library binding) | ISBN 9781647471033 (paperback) | ISBN
   9781647471125 (ebook)
Subjects: LCSH: Black mamba—Juvenile literature.
Classification: LCC QL666.O64 B68 2021 (print) | LCC QL666.O64 (ebook) |
   DDC 597.96/42—dc23
LC record available at https://lccn.loc.gov/2020008977
LC ebook record available at https://lccn.loc.gov/2020008978

For more information, write to Bearport Publishing, 5357 Penn Avenue South, Minneapolis, MN 55419. Printed in the United States of America.

Contents

Danger on the Path

A young woman was walking toward her home in the African country of Kenya. She didn't see the snake on the path ahead. But she felt the bite and a burning pain. She had been bitten by a black mamba.

The woman was rushed to the hospital where doctors gave her **antivenin** to save her life. She was lucky. Not everyone survives an attack by a deadly black mamba.

A bite from a black mamba can kill a person in 30 minutes.

Home For a Killer

Black mambas live in western, eastern, and southern Africa. They are found in **savannas**, woodlands, forests, and rocky hills.

The deadly snakes hunt during the day. At night, they sleep in hollow trees or between cracks in rocks. Sometimes, they find shelter in old animal burrows or termite mounds.

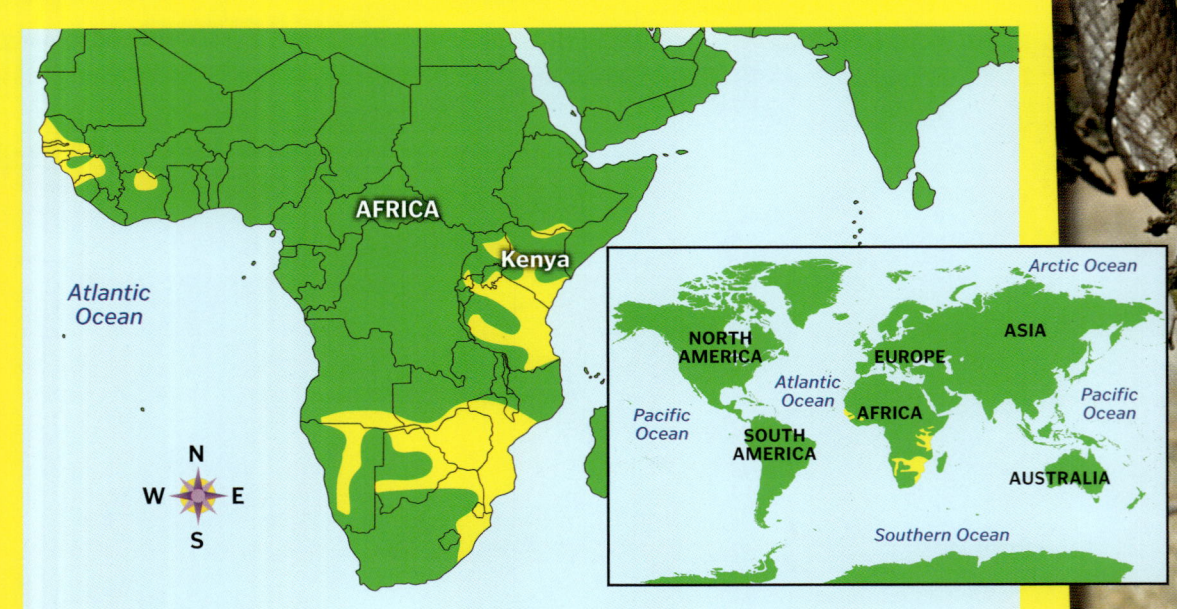

Where black mambas live

A black mamba often sleeps in the same place each night.

A termite mound

Long and Fast

At up to 14 feet (4.3 m) long, black mambas are Africa's longest venomous snakes. They are the fastest snakes in Africa, too. **Scales** on the snakes' bellies grip the ground and help these snakes zip along at up to 12 miles per hour (19 kph). That is faster than most people can run! Despite their name, black mambas are actually brown or gray with a pale belly.

The black mamba's name comes from the inky black color inside its mouth. The snake opens its mouth when it is **threatened**. You don't want to see the black on a black mamba!

A Deadly Bite

What makes the black mamba so dangerous? It's the mix of powerful **venom** and sharp **fangs**. The snake's 0.5-inch (6.5 cm) fangs are hollow, like a doctor's needle. And they deliver a deadly bite. A single bite has enough venom to kill at least 10 people!

Fleshy cover

Tongue

A black mamba's fangs stay inside fleshy covers until the snake is ready to bite.

On the Menu

A black mamba eats rodents, bats, and birds. It can also hunt small mammals such as hyraxes. The skilled **predator** bites its **prey**, lets it go, and follows. Before long, the venom has killed the victim. Then, it's time to eat! The black mamba swallows its prey whole. Strong acids in its stomach **digest** the meal.

A hyrax

A black mamba can open its mouth wide enough to swallow an animal that is four times as wide as the snake's head!

A fruit bat

Time To Mate

Black mambas **mate** in spring and summer. Male snakes fight over female snakes. The winner mates with her. Then, the female snake finds a warm, damp place for her eggs. She lays between 6 and 25 eggs and leaves the eggs. She never sees her young. About three months later, the eggs hatch.

Two males fighting

The female usually lays her eggs in a termite mound. The warmth in the mound helps the young snakes grow inside their eggs.

A black mamba hatching from its egg.

Deadly Babies

Black mamba babies are around 20 in (50 cm) long when they hatch. And they are deadly from the start! Newly hatched black mambas already have strong venom. They can hunt and catch prey right away. The young snakes grow quickly and can be 6.5 ft (2 m) long by the end of their first year.

A young black mamba

Old skin

Black mambas **shed** their skin from time to time as they grow. Young black mambas shed their skin more often than old ones.

Stay Away!

Despite their powerful venom, black mambas still have some predators. Mongooses catch young black mambas and eat black mamba eggs. Eagles and secretary birds can catch and kill the adult snakes.

A black mamba prefers to slither away if it is threatened. If the snake can't escape, it tries to scare its attacker. It lifts the front of its body off the ground, opens its mouth, and hisses. If this doesn't work, the snake strikes!

A secretary bird kills with a powerful kick to the snake's head. Then, the bird swallows the snake whole.

A secretary bird

A snake

Milking a Snake

Before the early 1960s, anyone who was bitten by a black mamba would die. Now, a medicine called antivenin can save the person's life. Antivenin is made from the snake's venom, which is collected by a process called milking. An expert holds the snake carefully as the creature bites onto something. The venom comes out of the fangs and is collected. It is turned into antivenin. Then, the snake venom that can be so deadly can also save lives!

Black mambas usually stay away from people. But people are starting to live closer to the black mambas' **habitat**. That makes it more likely for people to be bitten.

Black Mamba Facts

The black mamba's scientific name is *Dendroaspis polylepis*. *Dendroaspis* means poisonous tree snake. *Polylepis* means having many scales.

There are four different **species** of mambas. The other three are the western green mamba, the eastern green mamba, and Jameson's mamba. All mambas have powerful venom, but the black mamba is the most venomous of the four.

A western green mamba

Black mambas can live up to 11 years in the wild.

Black mambas can zip along holding a third of their bodies off the ground with their heads about 4 ft (1.2 m) in the air.

Black mambas belong to a family of venomous snakes called the elapid family. King cobras also belong to this family.

Glossary

antivenin a medicine that is used to treat snake bites

digest to break down food inside the body

fangs pointed teeth

habitat a place in nature where an animal lives

mate to come together to have young

predator an animal that hunts and kills other animals for food

prey animals that are hunted and killed by other animals

savannas grasslands with very few trees

scales tough, waterproof coverings on a snake's body

shed to lose a layer

species groups that animals are divided into, according to similar characteristics

threatened being in immediate danger

venom poison from a snake that is injected through hollow fangs; if something is venomous, it has venom

Index

Read More

Franchino, Vicky. *Black Mambas (Nature's Children).* New York: Children's Press (2016).

Owings, Lisa. *The Black Mamba (Pilot. Nature's Deadliest).* Minneapolis: Bellwether Media (2013).

Pope, Kristen. *On the Hunt with Black Mambas (On the Hunt with Animal Predators).* Mankatao, MN: Child's World (2016).

Learn More Online

1. Go to **factsurfer.com**
2. Enter "**Black Mamba**" into the search box.
3. Click on the cover of the book to see a list of websites.

About the Author

Craig Boutland has written many science, history, and nature books for children. He divides his time between London and the English countryside.